Sustainable Management of Solar PV Panel Waste

A Circular Approach to Solar PV Waste

Published 2025 by River Publishers
River Publishers
Broagervej 10, 9260 Gistrup, Denmark
www.riverpublishers.com

Distributed exclusively by Routledge
605 Third Avenue, New York, NY 10017, USA
4 Park Square, Milton Park, Abingdon, Oxon OX14 4RN

Sustainable Management of Solar PV Panel Waste: A Circular Approach to Solar PV Waste / Virendra Kumar, Nishant Kumar Singh, Dan Bhadur Pal and Surendra Kumar Yadav.

Routledge is an imprint of the Taylor & Francis Group, an informa business

ISBN 978-87-4380-085-9 (paperback)
ISBN 978-87-4380-087-3 (online)
ISBN 978-8-743-80086-6 (ebook master)

Sustainable Management of Solar PV Panel Waste

A Circular Approach to Solar PV Waste

Virendra Kumar
Madan Mohan Malaviya University of Technology, Gorakhpur, India

Nishant Kumar Singh
Harcourt Butler Technical University, Kanpur, India

Dan Bhadur Pal
Harcourt Butler Technical University, Kanpur, India

Surendra Kumar Yadav
K R Mangalam University Gurugram, India

River Publishers

Routledge
Taylor & Francis Group
NEW YORK AND LONDON

Contents

Preface

The global transition toward renewable energy has witnessed an unprecedented rise in the adoption of solar photovoltaic (PV) technology. As a cornerstone of sustainable energy solutions, solar PV panels have revolutionized the way we harness clean energy. However, alongside this rapid growth lies a critical challenge: managing the vast amounts of waste generated by end-of-life PV panels. The need for sustainable waste management strategies is paramount to ensure that the environmental gains achieved through solar energy are not offset by the unintended consequences of improper disposal and material wastage.

This book, titled *Sustainable Management of Solar PV Panel Waste*, aims to address this emerging challenge by providing a comprehensive exploration of the subject. The chapters are carefully structured to offer readers a holistic understanding of the life cycle of solar PV panels and the intricacies of managing their waste in an environmentally and economically sustainable manner.

Chapter 1 sets the stage by introducing the background, significance, and objectives of this study. It outlines the scope of the work and provides a roadmap for the organization of the subsequent chapters.

Chapter 2 delves into projections of PV panel waste, analyzing the rapid growth of the solar PV market and presenting models to predict future waste generation. This chapter highlights the urgency of developing effective waste management systems to accommodate the anticipated surge in end-of-life PV panels.

Chapter 3 examines the composition of PV panels and classifies waste based on hazard levels and regulatory frameworks. Understanding the materials and processes involved in PV manufacturing is essential for identifying viable recycling and disposal options.

Chapter 4 explores various waste management strategies, including reduction, reuse, and recycling, while emphasizing the potential of a circular economy. This chapter sheds light on innovative practices that align with global sustainability goals.

Chapter 5 provides a national perspective through case studies from different countries. By examining regulatory landscapes and industry actions, it identifies best practices and lessons that can inform global efforts to manage PV panel waste effectively.

Chapter 6 discusses value creation from end-of-life PV panels, focusing on technological innovations, resource recovery, and socioeconomic benefits. It underscores the potential of transforming waste management challenges into opportunities for economic and environmental gains.

Finally, Chapter 7 concludes the book by summarizing the key findings, offering recommendations for industry stakeholders, and suggesting avenues for future research and policy development.

This book is a valuable resource for researchers, policymakers, industry professionals, and environmental advocates who are committed to fostering sustainable practices in the renewable energy sector. By addressing the multifaceted aspects of PV panel waste management, it seeks to contribute to a cleaner, greener future.

The journey toward sustainable energy solutions does not end with the adoption of renewable technologies. It extends to how we manage their life cycle impacts. I hope this book serves as a catalyst for meaningful discussions and actions that ensure the longevity of solar energy as a truly sustainable resource.

About the Authors

Dr. Virendra Kumar is an Associate Professor in the Department of Mechanical Engineering at Madan Mohan Malaviya University of Technology, Gorakhpur, India. He has a Ph.D. in Thermal Engineering from the prestigious Indian Institute of Technology, Delhi, India and has over nine years of experience in teaching and research. He is an author of over 50 research articles and four books.

Dr. Nishant Kumar Singh is an Associate Professor at Harcourt Butler Technical University, Kanpur, India. He holds a Ph.D. from the Indian Institute of Technology (Dhanbad), India. Dr. Singh has published over 110 research articles in well-known international journals. He has also written five books with international publishers like CRC Press and Nova Publishers. His research interests include tribology, composites, micro-manufacturing, and non-conventional machining processes.

Dr. Dan Bahadur Pal is currently working as an Assistant Professor in the Department of Chemical Engineering at Harcourt Butler Technical University, Kanpur, India. He completed his Ph.D. in 2017 in Chemical Engineering from the Indian Institute of Technology (BHU) Varanasi, Uttar Pradesh, India. He has published 82 papers in internationally refereed journals, six books, and thirty book chapters.

Dr. Surendra Kumar Yadav is an Assistant Professor in the Department of Mechanical Engineering at K R Mangalam University in Gurugram, India. He earned his Master's (M.Tech) and doctoral (Ph.D.) degrees from the National Institute of Technology in Silchar, India. He has published more than 21 articles in reputed international peer-reviewed journals.

CHAPTER 1

Introduction

1.1 Background and Significance

The fast growth of solar PV technology has resulted in considerable movement toward sustainable energy generation. In the last few decades, solar PV systems have emerged as a critical component in the global transition from fossil fuels to alternative energy sources. Moving to a low-carbon economy is propelled by the need for an urgent response to climate change, the avoidance of greenhouse gas emissions, and better energy security. Solar PV panels, which are designed to convert direct sunlight into electricity, have been one of the leading sources of energy revolution as a clean energy solution that offers a sustainable and ever-more cost-effective alternative to traditional energy solutions.

Now, as more and more solar PV systems are deployed, there is a new problem: PV panel waste. The lifespan of solar panels ranges from 25 to 30 years and the first generation of panels is at the end of its life cycle. According to the International Renewable Energy Agency (IRENA), global PV panel waste could rise to 78 million tons by 2050. This upcoming tsunami of PV panel waste is a real concern for the environment; however, it also provides a chance to practice sustainable waste processes and extract valuable metals. The balancing of the impacts of PV panel waste management is shown in Figure 1.1.

PV panel waste must be processed correctly. Insufficient disposal may lead to environmental contamination, resource depletion, and lost economic opportunities. The management of PV panels has paramount importance in the current scenario. On the other hand, good waste management can minimize the environmental impact, conserve resources, and produce new economic opportunities through recycling and reusing.

Figure 1.1. Balancing the impacts of PV panel waste management.

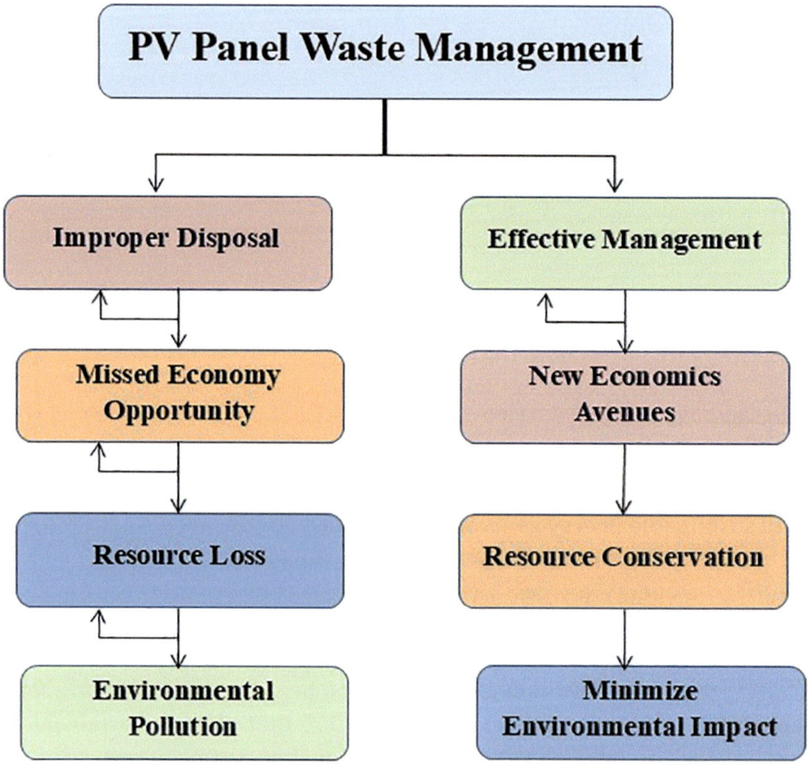

1.2 Objectives of the Study

The overall aim of this book focuses on the sustainable management of solar PV panel waste. In particular, the study seeks to:

(1) Learn about the planned growth of solar PV installations and the resulting generation of PV panel waste globally.
(2) Know how PV panels are made and how to classify the different types of waste.
(3) Investigate the practices of varying kinds of waste handling like reduction, reuse, and recycling. Determine whether or not you think it will work or work well.

(4) Be aware of laws and regulations concerning PV panel waste management.
(5) Best practices and lessons learned from countries with mature PV waste management systems and programs.
(6) Explore technological innovations and economic opportunities for the value creation from end-of life PV panels.

1.3 Scope and Limitations

This book deals with a number of issues related to the sustainable management of solar PV panel waste from a global perspective including in-depth case studies from key solar markets: The European Union, Germany, the United Kingdom, Japan, the United States, China, and India. General deal: the book is a full life cycle of the PV panels covering: manufacture, deployment, end-of-life disposal, and recycling. The case study, scope, and limitation of the sustainable management of solar PV panel waste are illustrated in Figure 1.2.

Figure 1.2. Sustainable management of solar PV panel waste.

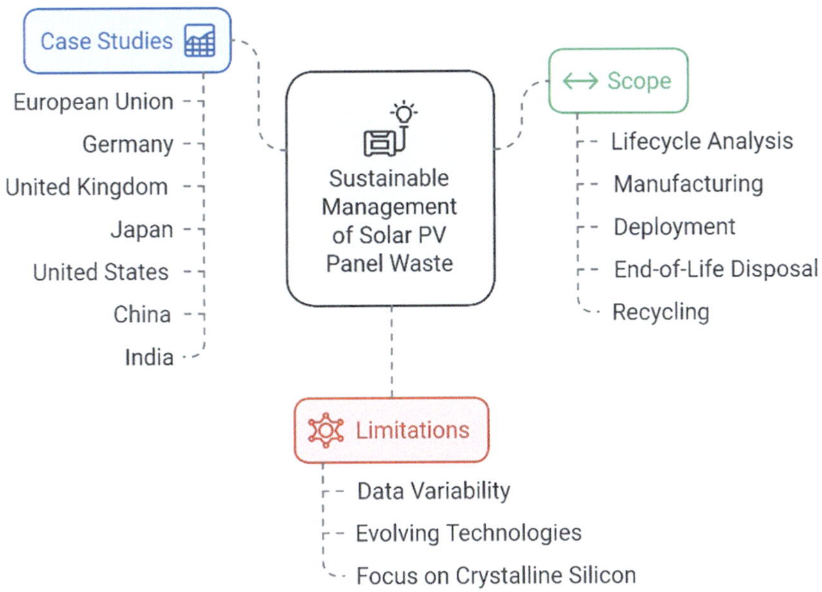

The study has some limitations. First, the availability and quality of data on PV panel waste vary significantly across regions, which may affect the accuracy of waste projections and impact assessments. Second, while the book aims to cover emerging technologies and innovations in waste management, the rapidly evolving nature of this field means that some developments may not be fully captured. Lastly, the focus is primarily on crystalline silicon PV panels, which constitute the majority of the market, but other types of panels, such as thin-film technologies, are also considered where relevant.

CHAPTER

2

Solar PV Panel Waste Projections

2.1 Global Solar PV Growth

2.1.1 Historical trends

In the last few decades, the solar photovoltaic (PV) industry has evolved through notable phases and embraced rapid technological industrial progress, its evaluation phases are outlined in Figure 2.1. With its origins rooted in the 1950s, when the first practical PV cell was developed by Bell Labs and ranging to sizes approaching megawatts around the world today, the ascent of solar PV has truly been exponential.

The high cost and low efficiencies of those early years meant that solar PV technology was practically a no-show on Earth during its first years. However, large investments in research and development, and also favorable policies from many governments, brought about enormous technological progress. By the late 1990s and early 2000s, however, the efficiency of PV cells had increased dramatically and costs were coming down, although price-performance still did not match that of other electrical energy sources. During the same period, feed-in tariffs (FITs) and other incentive mechanisms introduced in countries including Germany and Japan were instrumental in generating first market growth.

Figure 2.1. The PV panel industry evolution.

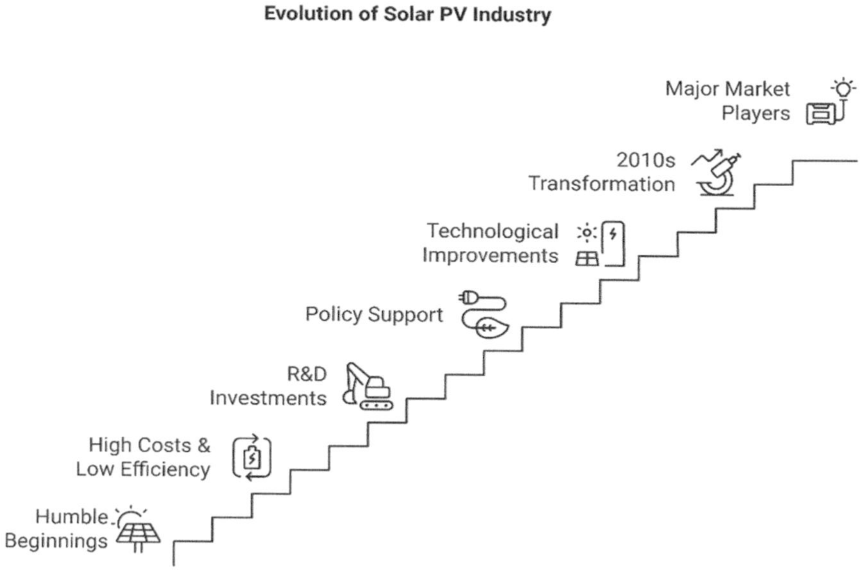

The solar PV industry underwent a transformative decade in the 2010s. Thanks to economies of scale, new manufacturing processes, and stagnating competition, the price of solar PV was falling through the floor. The price drop, in addition to rising environmental concerns and the necessity for clean energy sources, caused a boom around the world in installations. The major countries in the solar PV sector, including China, the United States, and India, became responsible for substantially all of the global capacity additions.

2.1.2 Current market dynamics

The global solar photovoltaic (PV) market has experienced remarkable growth in recent years. By the end of 2024, the cumulative installed PV capacity worldwide surpassed 2 terawatts (TW), doubling from 1 TW in 2022.

This rapid expansion is expected to continue, with projections indicating that global solar installations will reach approximately 580 gigawatts (GW) in 2025. China has been a significant contributor to this growth, leading the world in solar farm capacity with 386,875 megawatts (MW) as of June 2024.

The International Energy Agency (IEA) anticipates that the installed capacity of both utility-scale and distributed PV sectors will almost quadruple between 2023 and 2030, underscoring the increasing importance of solar PV in the global energy mix. This growth is supported by the factors mentioned in Figure 2.2.

Figure 2.2. Factors influencing solar PV panel growth.

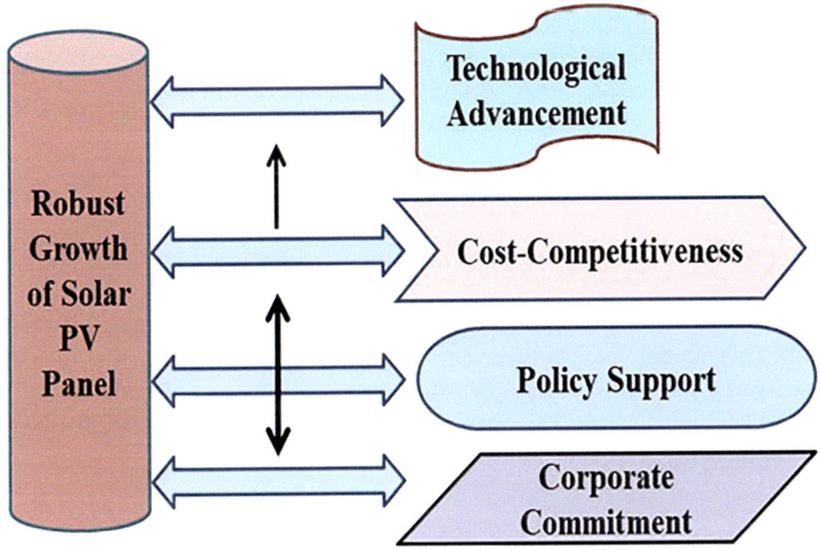

- *Technological improvements*: ongoing advances in PV cell performance, the introduction of bifacial panels, and new approaches in system design have increased the competitiveness of solar PV.
- *Cost competitiveness*: solar PV has reached grid parity in numerous countries, allowing it to compete cost-effectively without subsidies against conventional energy sources.
- *Policy incentives*: in many parts of the world, solar PV is supported by RPS/fits/tax credits/ auction mechanisms available.
- *Corporate commitments*: more and more companies are committing to targets of 100% renewable energy, spurring demand for solar installations.

2.1.3 Future projections

The future of solar PV looks bright as well. The multitude of industry forecasts indicate the world installed capacity of solar PV could surpass 3000 GW by 2030. There are a few factors behind these rosy projections:

- *Net zero emissions commitment*: over 120 nations have pledged to net-zero emissions by midcentury, and solar PV is a key element of their renewable energy commitments. In addition, in cases where construction and maintenance of electricity grids are not possible, the use of solar PV can be an alternative, promoting access to energy on a global scale.
- *Technological advances*: continued exploration of new materials – like perovskites – and advanced manufacturing processes will only improve efficiency and decrease solar PV costs.
- *Investment trends*: there is a growing trend of investments in renewable energy infrastructure by both public and private sectors, bolstering the deployment of solar PV projects.

2.2 PV Panel Waste Model

2.2.1 Development methodology

A reliable waste model needs to be developed for projecting future PV panel waste. This model must incorporate several factors, including the installed capacity of PV systems, average lifespan of PV panel's power, degradation rates, and replacement trends. This means that the methodological way of creating this model is broken into a few key points as illustrated in Figure 2.3.

Figure 2.3. PV panel waste model.

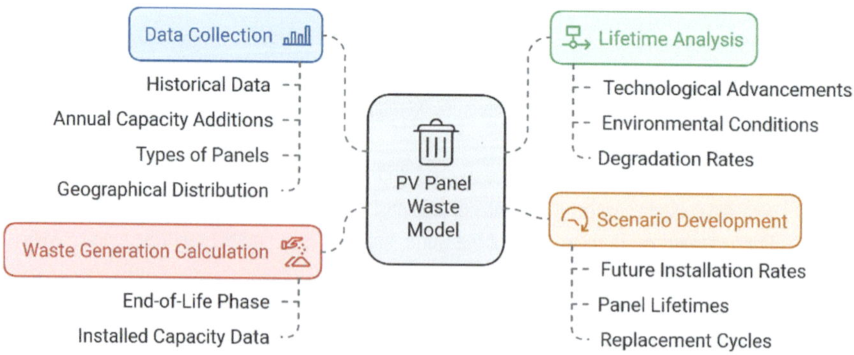

Data collection: this involves gathering historical data on PV installations such as annual capacity additions, types of panels used, and geographical distribution.

Scenario development: develop a range of scenarios knitted with varying assumptions regarding future installation rates, the lifetimes of panels, and replacement cycles.

Waste generation calculation: estimate annual waste generation of the developed scenarios with installed capacity and end-of-life phase of PV panels.

2.2.2 Key variables and assumptions

The accuracy of the PV panel waste model relies on several key parameters and assumptions:

- *Lifespan of PV panels*: the average lifespan is generally considered to be ~25–30 years. Nevertheless, the actual lifespan might differ depending on the quality of the panels, the installation techniques used, and the conditions they are exposed to.
- *Degradation rate*: panels deteriorate over time and this affects their efficiency and lifespan. The average annual degradation is around 0.5% to 1%.
- *Replacement trends*: improving efficiency might lead to replacing certain panels before their useful life is up, as they might suffer damage.
- *Technological improvements*: enhancements in panel durability and efficiency can stretch their lifetime and impact waste predictions.

2.2.3 Sensitivity analysis

A sensitivity analysis of waste characterization variables should be conducted. They vary panel lifetime, degradation rates, and installation growth rates to estimate their turns out of overall uncertainty in waste generation estimates. Through a scenario-tapping mechanism, we can understand the ones shaping PV panel waste the most, and cleaner and more accurate projections can be derived for risk assessment.

2.3 PV Panel Waste Projections

2.3.1 Regional breakdowns

Generation of waste from PV panels will differ widely between regions, mirroring differences in installation rates, types of panels being used, and their lifespans. Here is a regional breakdown of projected PV panel waste:

- *Europe*: Europe was the early leader in solar PV deployment, with huge amounts of capacity installed over two decades. Consequently, most of the elderly PV modules at the end of their lifecycle, especially older installations in Europe, will create a significant amount of PV panel waste.
- *Asia-Pacific*: the highest growth in solar PV installations is seen in the Asia-Pacific region, dominated by China. As a result, the region is expected to contribute the highest proportion of global PV panel waste over the coming decades.
- *North America*: the United States and Canada have also experienced significant growth in solar PV capacity. Waste generation in this region will reflect the maturity of the market and the replacement of older panels.
- *Other regions*: emerging markets in Latin America, Africa, and the Middle East are expected to contribute to PV panel waste generation as their solar sectors expand.

2.3.2 Sectoral impact assessment

The impact of PV panel waste extends beyond environmental considerations and influences various sectors as shown in Figure 2.4.

Figure 2.4. Impact of PV panel waste on different sectors.

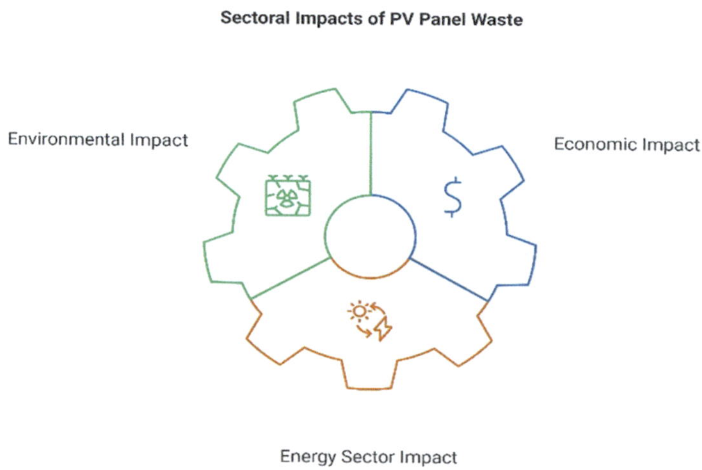

- *In the environment*: improper disposal of photovoltaic (PV) panels can release toxic materials, such as cadmium and lead, into the atmosphere. Reduction of these risks is only possible if the waste management practices are effective.

- *Economic*: recycling and reusing PV panels provides economic benefits such as the recovery of critical and valuable materials like silicon, silver, and aluminum.
- *Energy sector*: the sustainable and credible future of the solar energy sector lies in the efficient management of PV panel waste. From thrift to rot, failure to tackle waste issues will erode support for renewables.

2.3.3 Environmental implications

Solar panels also have serious performance aspects in terms of environmental factors from their waste. Poorly managed disposal of panels, at the end of their life, can pollute soil and water and have adverse effects on human health and ecosystems. Furthermore, improper waste disposal leads to the wastage of valuable resources that could have been recycled and reused, contributing to the efficient use of materials.

This means developing and putting in place sustainable waste disposal strategies that focus on recycling such materials. We must be careful to responsibly recycle PV panels to minimize their environmental burden, ensuring we head toward a new sustainable, circular economy. Overall, the data referring to PV panel waste highlight the critical situation concerning proper waste management. The sector must come together to address the challenges posed by both PV panel waste. The upcoming chapters elaborate on the attributes of energy waste and an overview of waste classification with a classification structure to substantially highlight the different stages of waste management options and the crucial timelines for PV panels, thus providing an example of a sustainable PV panel waste management roadmap.

CHAPTER

3

PV Panel Composition and Waste Classification

3.1 Photovoltaic (PV) Panel Composition

We can see that photovoltaic (PV) panels are made of many layers of varying materials, each with an essential role in helping to transform the energy of sunlight into electricity. Below is a breakdown of PV cell components and functionalities.

3.1.1 Semiconductor materials

- *Silicon (Si)*: the most common type of solar cell is silicon-based, particularly, monocrystalline and polycrystalline silicon. The silicon-based PV cell compositions and cell structure are shown in Table 3.1 and Figure 3.1, respectively.
- *Monocrystalline silicon (Mono-Si)*: more efficient and costly, but delivers high power outputs (efficiency ~20%–22%).
- *Polycrystalline silicon (Poly-Si)*: less expensive and slightly less efficient than monocrystalline (efficiency ~15%–18%).
- *Amorphous silicon (a-Si)*: a thin-film technology that is less efficient (~10%–12%) but cheaper and flexible, often used in portable applications.
- *Cadmium telluride (CdTe)*: CdTe is a leading thin-film technology and the second most common PV material. CdTe modules are cheaper to produce but raise environmental concerns due to cadmium toxicity.
- *Copper indium gallium selenide (CIGS)*: CIGS cells offer better efficiency than other thin-film technologies (~13%–15%) but are still less common due to high production costs.
- *Perovskite*: perovskite materials, an emerging technology, promise high efficiency (over 25%) and low-cost production, though concerns remain regarding their stability and use of toxic lead.

3.1.2 Conductive materials

- *Silver (Ag)*: used in the grid lines of PV cells to collect and conduct electrons. Silver is chosen for its high electrical conductivity, although cost considerations push research into alternatives like copper.
- *Aluminum (Al)*: commonly used in the frame and back-contact layer for electrical conductivity and structural integrity.

3.1.3 Encapsulation and backing materials

- *Encapsulation layer*: typically made of *ethylene vinyl acetate (EVA),* this layer surrounds and protects the delicate solar cells from moisture and mechanical damage.
- *Backsheet material*: generally made of *polyvinyl fluoride (PVF)* or other durable polymers, the backsheet provides protection from the environment while offering electrical insulation.

3.1.4 Glass layer

- The top layer of a PV panel is typically made of tempered, low-iron glass that allows maximum sunlight transmission while protecting the underlying cells from environmental hazards like hail, dirt, and extreme weather.

3.1.5 Frame

- Most PV panels are housed in an aluminum frame, which adds strength and allows for easier mounting and integration into solar arrays.

Table 3.1. Composition of typical silicon-based photovoltaic cells.

Component	Material	Functions
Semiconductor	Monocrystalline or Poly-Si	Converts sunlight into electricity via the photovoltaic effect
Front contact	Silver (Ag)	Collects electrons and transports them to an external circuit
Back contact	Aluminum (Al)	Provides an electrical pathway for electron return
Encapsulation	EVA	Protects cells from physical and environmental damage
Glass cover	Tempered glass	Protects cells while allowing maximum sunlight penetration
Frame	Aluminum	Provides structural support and protection from environmental stress
Backsheet	Polyvinyl fluoride (PVF)	Protects the backside of the panel from moisture and mechanical damage

Figure 3.1. Cross-sectional diagram of a silicon-based photovoltaic cell structure.

3.2 Waste Classification

PV panels, like all electronic devices, generate waste at every stage of their lifecycle from raw material extraction to end-of-life disposal. The classification of this waste is crucial for its proper management and recycling.

3.2.1 Types of waste generated by PV cells

3.2.1.1 Manufacturing waste

- *Silicon tetrachloride ($SiCl_4$)*: produced during the manufacturing of polysilicon, $SiCl_4$ is highly reactive with water and can produce toxic gases (hydrochloric acid). Proper treatment and recycling of $SiCl_4$ are essential to avoid environmental hazards.
- *Hydrofluoric acid (HF)*: used for cleaning and texturizing silicon wafers, HF is highly corrosive and poses risks to both human health and the environment.

3.2.1.2 End-of-life waste

- *Shattered panels*: PV panels may shatter in the process of use or installation, and if not properly handled, they can release harmful substances such as lead, cadmium, and other toxins into the environment.

- *Toxic materials in panels*: cadmium telluride (CdTe) and CIGS cells contain cadmium and selenium, both of which are considered hazardous materials.

3.2.1.3 Decommissioning and waste recycling

- *EVA and PVB encapsulation*: these materials decay slowly and can be a source of microplastics if not properly recycled
- *Glass and aluminum*: though these elements are easier to recycle, they still need to be collected and sorted carefully to make sure it is not mixed with harmful materials.

Table 3.2. Classification of waste from PV cells.

Waste type	Source	Hazardous components	Treatment
Silicon tetrachloride	Polysilicon manufacturing	Hydrochloric acid, reactive with water	Neutralization and recycling
Hydrofluoric acid	Wafer cleaning	Corrosive, toxic fumes	Chemical neutralization, strict handling protocols
Broken panels	Operational damage	Lead, cadmium	Safe handling, recycling, and disposal
Encapsulation material	Decommissioned panels	EVA, PVB, microplastics	Mechanical separation, pyrolysis
Glass and aluminum	Decommissioning	None (contaminated with trace toxic substances)	Recycling, glass-to-glass technology

3.2.2 Hazardous effects of PV waste

The classification of waste from PV cells is illustrated in Table 3.2. Inadequate disposal of PV panel waste can pose serious environmental and health hazards, primarily because of the presence of toxic materials such as cadmium, lead, and selenium. Specific examples include:

- *Lead (Pb)*: when used in soldering connections in PV panels, lead can leach out into the soil and water supply, causing neurological and developmental damage, especially in children.
- *Cadmium (Cd)*: cadmium is present in CdTe cells and is a known carcinogen that can persist in soil and water, causing severe health consequences if released into the environment.

These risks must be managed, which is why the importance of good regulatory frameworks such as the WEEE Directive and extended producer responsibility (EPR) is paramount.

3.2.3 Waste electrical and electronic equipment (WEEE) directive

The European Union's Waste Electrical and Electronic Equipment (WEEE) Directive establishes rules in the collection, recycling, and recovery of electrical waste, incorporating PV panels. It requires producers to bear the cost of collecting and treating the waste produced by their products.

3.2.4 Extended producer responsibility (EPR)

Hard regulations: extended producer responsibility (EPR) mandates manufacturers to manage the complete lifecycle of their products, from production to postconsumption waste management. Extended producer responsibility (EPR) policies encourage product design for recycling and maker responsibility for recycling costs.

3.3 Life Cycle Assessment (LCA) of PV Cells

Life cycle assessment (LCA) is a method used to evaluate the environmental impacts associated with all stages of a product's life. For PV cells, the key phases are illustrated in Figure 3.2.

Figure 3.2. Life cycle assessment of PV cells.

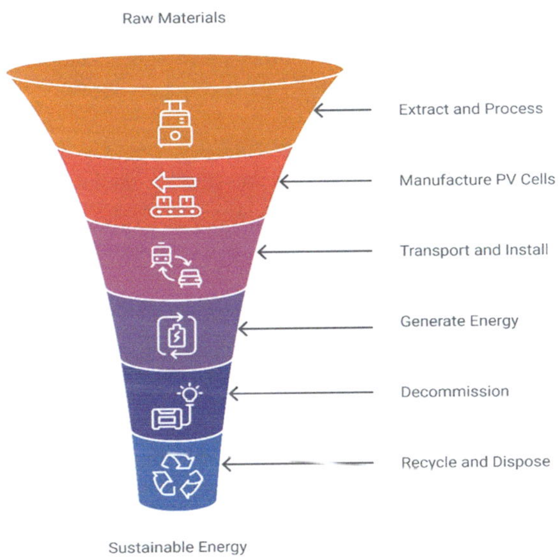

3.4 Raw Material Extraction and Processing

- *Input:* silicon (Si), silver (Ag), aluminum (Al), glass, copper (Cu).
- *Process:* mining, refining, and chemical processing to produce silicon wafers, glass sheets, and metal components.
- *Environmental impact:* high energy consumption, CO_2 emissions, silicon tetrachloride waste, and other hazardous materials (cadmium in CdTe cells).
- *Output:* refined materials for manufacturing (silicon wafers, conductive materials, glass, aluminum frames).

3.5 Manufacturing of PV Cells and Modules

- *Input:* refined materials (silicon wafers, silver, aluminum, glass), energy, chemicals (HF, $SiCl_4$)
- *Process:*
    - Creation of semiconductor layers, deposition of conductive materials, lamination of PV cells between glass and backsheet materials, and assembly into modules.
    - Chemical processes used for cleaning wafers, texturizing surfaces, and doping semiconductors.
- *Environmental impact:* energy-intensive process with significant emissions (CO_2, HF, $SiCl_4$), material waste.
- *Output:* PV panels ready for installation, manufacturing waste (scrap wafers, toxic chemicals).

3.6 Transport and Installation

- *Input:* finished PV panels, transportation fuel, installation tools, and materials.
- *Process:* shipping of PV panels from the manufacturing site to the installation location. Installation on rooftops or ground-mounted arrays.
- *Environmental impact:* emissions from transportation, minimal environmental impact during the installation phase.
- *Output:* installed PV systems ready to generate electricity.

3.7 Use Phase (Energy Generation)

- *Input:* solar energy (sunlight).
- *Process:* photovoltaic conversion of sunlight into electricity, maintenance of PV panels over their operational life (25–30 years).

- *Environmental impact:* virtually no direct emissions or waste during the use phase. Contribution to reducing CO_2 emissions by offsetting fossil fuel-based electricity.
- *Output:* clean electricity, minimal maintenance waste (e.g., replaced components).

3.8 Decommissioning and End-of-Life (EoL)

- *Input:* decommissioned PV panels.
- *Process:* dismantling, transportation to recycling facilities, waste sorting, and treatment.
- *Environmental impact:* potential release of hazardous materials (lead, cadmium, selenium) if panels are improperly disposed of. If recycled, recovery of valuable materials (glass, aluminum, silicon).
- *Output:* recycled materials (glass, metals), and hazardous waste if not managed properly.

3.9 Recycling

- *Input:* decommissioned PV panels, energy for recycling processes.
- *Process:*
    - Separation of materials through mechanical or chemical means (e.g., crushing, pyrolysis).
    - Recovery of high-purity silicon, glass, and metals (aluminum, silver).
- *Environmental impact:* reduces the need for raw material extraction, and lowers energy use compared to virgin material production, but still involves emissions from recycling processes.
- *Output:* recycled glass, silicon, and metals; residual waste (encapsulant plastics, toxic substances).

3.10 Waste Disposal (Nonrecyclable Waste)

- *Input:* encapsulants (EVA, PVB), degraded or hazardous materials (lead, cadmium)
- *Process:* landfilling or incineration of nonrecyclable materials.
- *Environmental impact:* potential contamination of soil and water from hazardous substances, and microplastic pollution from degraded encapsulants.
- *Output:* landfill waste, toxic byproducts from incineration.

The LCA indicators for PV cells have significant importance in PV panel management, the LCA indicators for PV cells are outlined in Table 3.3.

Table 3.3. LCA indicators for PV cells.

Indicator	Unit	Typical value for PV panels
Global warming potential (GWP)	kg CO_2 eq/kWh	30–50
Energy payback time (EPBT)	years	2–4
Material recovery rate	%	85%–95% (for glass and aluminum)
Waste generation	kg/kW installed	50–60

CHAPTER

4

PV Panel Waste Management Options

Photovoltaic (PV) panels have become the cornerstone of sustainable energy solutions following the renewable energy boom. However, while so much can be done in using these panels to generate power that is good for the environment, a lifecycle will end one way or another. As a result, the disposal of panels has become a serious issue. The role of this policy was to understand how national PV waste policy can increase waste recycling efforts.

4.1 Waste Management Principles for PV Panels

An unprecedented increase in solar PV installation is observed globally due to solar photovoltaic (PV) technology commercialization by 2023. However, such explosive growth presents an issue that has to be tackled – solar PV panel waste. Data training is available until October 2023. The increasing deployment of solar energy systems has raised concerns regarding sustainable waste management for end-of-life (EoL) panels. This part helps to formulate the main waste management principles for the design of PV panels that are vital to reducing the ecological burden of solar technology. Such principles are typically grouped in the "3Rs," Reduce, Reuse, and Recycle, along with a growing emphasis on circular economy initiatives and technology-based solutions (refer Figure 4.1)

Figure 4.1. Waste management principles for the design of PV panels.

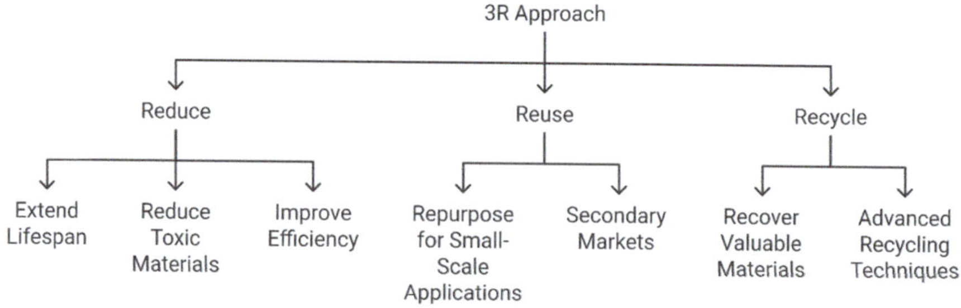

4.1.1 Reduce, reuse, and recycle (3R)

The "3R" approach is a fundamental waste management strategy adaptable for PV panel applications. All of the above is crucial to the efficient handling of EoL solar panels:

Reduce: the best waste management solution is reducing waste at the source. For PV panels, this might include innovations that extend their lifespan, minimize the amount of toxic materials used in their production, or increase efficiency in a way that reduces the number of panels needed to produce a given energy output. Strategies for reducing PV panel waste at the source are illustrated in Table 4.1. This goal can be addressed with significant effort in durable and long-lasting materials for PV panels.

Table 4.1. Strategies for reducing PV panel waste at the source.

Strategy	Description	Example
Material optimization	Use of less hazardous and longer-lasting materials	Thin-film PV panels using less cadmium and lead
Efficiency improvements	Increasing energy output per unit area	Bifacial solar panels capturing sunlight on both sides
Advanced manufacturing	Minimizing waste during the production phase	Automated precision manufacturing reducing material waste

Reuse: reusing PV panels eliminates or reduces waste and minimizes the life of materials used for solar technology. For example, PV panels that no longer perform up to the energy generation performance thresholds in one application scenario may still be utilized for other applications, such as decentralized energy applications (typically at lower generation scales), remote mini-grid generation, or secondary markets in lower-income nations.

Recycle: environmental impact avoidance at the highest possible level is to apply the 3R (reduce, reuse, and recycle) model, wherein the lower the position on the pyramid, the better the environmental impact of material consumption growing by orders of magnitude for materials used in PV panels (glass, aluminum, semiconductors, etc.), therefore, good recycling methods are capable to take value from the obsolete panels while reducing the demand for new raw materials and destructive extraction operations. Advanced recycling methods (e.g. suffering method and chemical recycling) can develop waste PV cell recycling and restore rare earth elements/ semiconductors.

4.1.2 Circular economy approaches

The circular economy includes the question of what to do with solar PV waste. The circular economy aims to design out waste and pollution, keep products and materials in use, and regenerate natural systems. The key circular economy strategies for PV panels regarding PV panel waste are illustrated in Table 4.2, the circular economy prolongs the life cycle of solar product systems by encouraging product redesign, reuse, and recycling.

Table 4.2. Key circular economy strategies for PV panels.

Circular economy strategy	Benefits	Example
Design for disassembly	Simplifies the recycling or refurbishment of panels	Modular panels where individual cells can be replaced
Remanufacturing	Extends the life of PV components	Refurbished inverters and modules
Product-as-a-service module	Reduces the need for ownership, encouraging reuse	Solar panel leasing and energy service companies (ESCOs)

The circular economy model can be applied to every stage of a PV panel's life cycle, from manufacturing to decommissioning (refer to Figure 4.2). For instance, solar PV companies can embrace design-for-disassembly principles, enabling simpler recycling or refurbishment during the EoL phase. Innovative modular designs allow modular PV modules which enable upgrading and replacing parts instead of replacing the entire panel.

Figure 4.2. The life cycle of a PV panel in a circular economy.

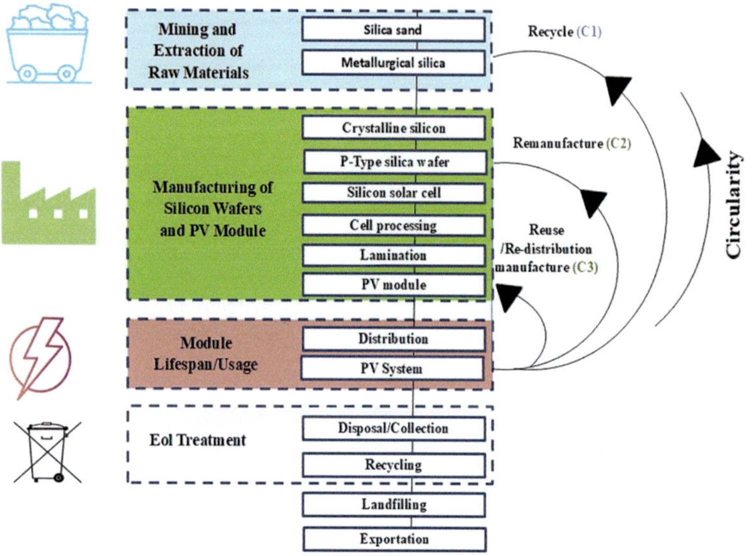

4.1.3 Technological innovations

Technology offers promising that is challenging to the huge amount of waste generated from PV panels. Materials science, manufacturing processes, and recycling technologies are rapidly advancing minimizing waste and recovering much more value from EoL panels.

Materials improvements: emerging materials, such as perovskite, have the potential to improve PV panel efficiencies while avoiding toxic materials like cadmium and lead. More recent engineering improvements in encapsulation materials have enhanced the disassembly of panels at the end of life for recycling.

Novel collection technologies: innovative concepts such as laser-based delamination and solvent-based recycling to recover materials from panels at end-of-life are already being explored to improve the efficiency of the collection process. This is used to recover high-purity silicon, rare earth elements, and semiconductor materials for recycling. The solar panel recycling processes adopted in most countries nowadays, along with key technological innovations in PV panel waste management, are illustrated in Figure 4.3 and Table 4.3, respectively.

Figure 4.3. Best practices for PV panel waste management in the EU.

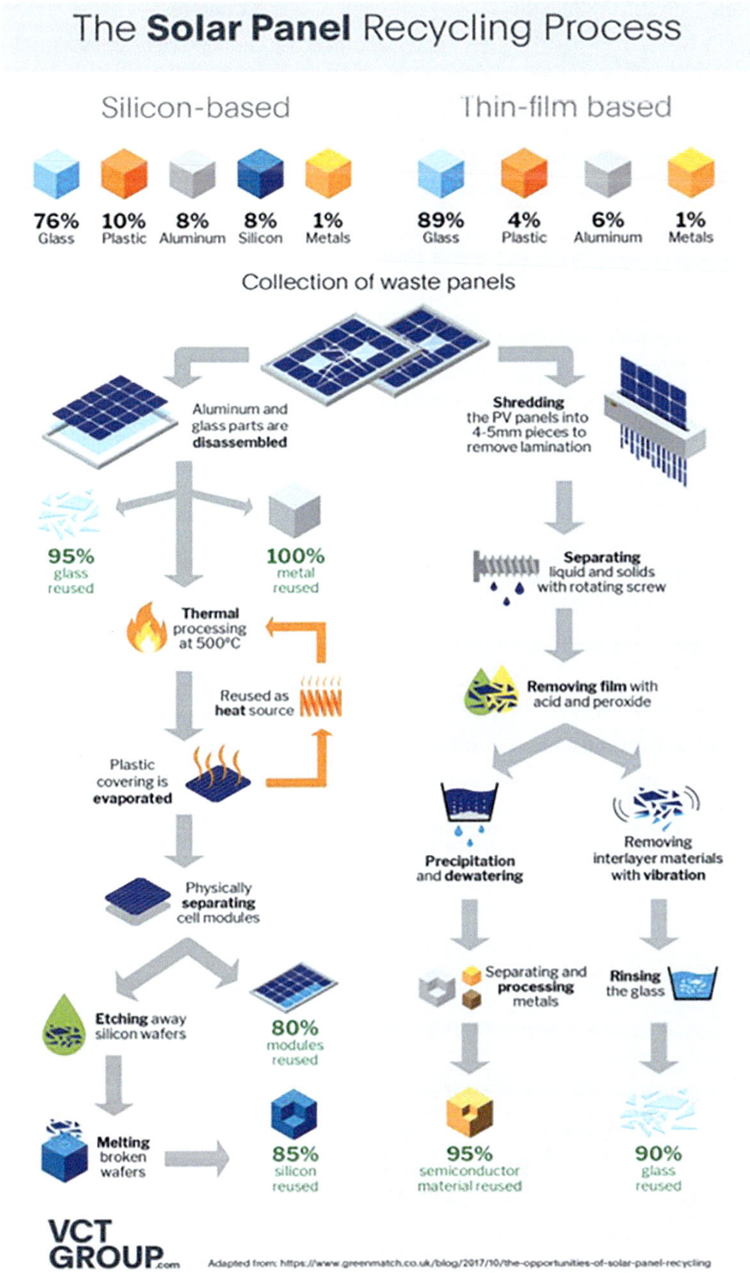

Table 4.3. Key technological innovations in PV panel waste management.

Innovation	Benefits	Examples
Perovskite PV panels	Higher efficiency, fewer toxic materials	Next-generation perovskite solar cells
Laser-based delamination	Efficient removal of solar cells from backing materials	Laser-cutting technology for panel recycling
Solvent-based recycling	Higher material recovery rates with lower energy consumption	Solvent methods to extract high-purity silicon and glass

4.2 Regulatory Approach: European Union

The EU has been the forerunner in establishing rules and frameworks for the proper disposal of PV panels. By doing so, the EU directives or policies enforced are tending toward the provision of adequate mechanisms for proper storage and treatment of solar PV waste in an environmentally sound manner. The next section provides an overview of these regulations along with case studies demonstrating their use.

4.2.1 Overview of EU directives

The EU regulatory basis for solar PV panel waste management is mainly shaped by the Waste Electrical and Electronic Equipment (WEEE) Directive (2012/19/EU). According to this directive, PV panels fall under the category of electronic waste, and manufacturers are obliged to pay for the collection, treatment, and recycling of decommissioned panels. This "extended producer responsibility" (EPR) is the ultimate in shifting the responsibility for waste management to the producers themselves and offering incentives to design and manufacture products in an environmentally responsible way.

Important WEEE Directive requirements relevant to PV panel waste include:

- *Producer responsibility:* manufacturers must manage and pay for the collection and recycling of PV panels.
- *Collection targets:* member states shall achieve collection and recycling targets for all electronic waste, including waste PV panels.
- *Recycling standards:* minimum efficiency rates for recycling recovery must be implemented, for example, recovering 85% of the weight of the PV panels.

The key provisions of the WEEE Directive for PV panel management to enhance the circular economy are illustrated in Table 4.4.

Table 4.4. Key provisions of the WEEE Directive for PV panels.

Provision	Requirement	Impact
Producer responsibility	Producers must manage collection, recycling, and disposal	Encourages circle design and waste minimization
Collection targets	Minimum collection rate of 65% for electronic waste	Ensure high rates of material recovery and recycling
Recycling efficiency targets	85% material recovery from decommissioned PV panels	Drives innovation in recycling technology and efficiency

4.2.2 Case studies

Several EU member states have implemented the WEEE Directive with notable success. This section presents two case studies:

- *Germany:* Germany has successfully implemented a well-functioning collection and recycling scheme for PV panels. It has also developed specialized recycling plants capable of recovering up to 95% of the metals, glass, and semiconductor materials from PV panels. Germany's experience highlights the crucial role of public–private partnerships in achieving high recycling rates.
- *France:* in France, producers of PV panels participate in a collective recycling system that pays for the collection and treatment of decommissioned panels. In France, it is of complaint that collaboration among manufacturers, recyclers, and government agencies is not as priority to execute waste management efficiently.

4.2.3 Key takeaways and best practices

Some lessons can be learned from the European experience:

- Extended producer responsibility (EPR) holds massive potential of making manufacturers responsible for the entire life cycle of their products (including waste).
- Investment in recycling infrastructure is essential for achieving high rates of material recovery. Governments should provide incentives for the development of advanced recycling technologies.
- Public awareness campaigns can significantly improve the collection rates for PV panel waste, ensuring that decommissioned panels do not end up in landfills.

CHAPTER

5

National Approaches to PV Waste Management

With the growth of the global solar photovoltaic (PV) industry, management of end-of-life (EoL) solar PV panels is progressively coming to be regarded as an upper priority. Diverse regulatory frameworks and industry practices have been adopted by individual countries in response to the increasing excess of PV waste. In this chapter, we cover the national strategies of Germany, the United Kingdom, Japan, the United States, China, and India to provide diverse lessons for the establishment of sustainable PV waste management systems. We look at the regulatory framework in these countries and industry-led initiatives and collaboration in this space, as well as hurdles, opportunities, and future potential policy directions.

5.1 Germany

5.1.1 Regulatory landscape

Germany has traditionally been a forerunner when it comes to environmental policies, including electronic waste disposal. Following the revision in 2012, solar panels are categorized as electronic waste under the European Union Waste Electrical and Electronic Equipment (WEEE) Directive and their EoL management must be conducted by their producers. The Waste Electrical and Electronic Equipment (WEEE) Directive was transposed into German national law with the German Electrical and Electronic Equipment Act (ElektroG). Under the framework, manufacturers, importers, and distributors of PV panels must organize and finance the collection and recycling of discarded panels.

In addition to the WEEE Directive, Germany also has a Circular Economy Act to promote the recovery of valuable materials from waste products, such as PV panels. Energy Resource Recovery, Recycling, Elimination Act of 2023, act, and to promote the efficient use of resources and decrease waste in landfills and circular economy principles such as photovoltaic panels. The right regulatory framework in Germany ensures that solar panels have to be managed as a priority waste stream, and the infrastructure needed for the collection, recycling, and proper disposal of photovoltaic (PV) waste has already been established.

5.1.2 Industry initiatives

Apart from regulatory frameworks, industry initiatives comprised a key component of Germany's photovoltaic waste management. PV Cycle, a European nonprofit, was created to help clients satisfy WEEE obligations. PV Cycle has collection points where solar panels can be dropped off for recycling throughout Germany. The organization also helps recycle, as well as dismantle panels, recovering materials like silicon, glass, and aluminum for reuse in other products.

PV cycle: success through example of how government regulations and industry collaboration can indeed work in harmony for sustainability in the solar space. The recovery rates for PV materials in Germany are one of the highest in Europe, recovering an average of 85%–90% of the materials in each panel that is recycled.

5.1.3 Lessons for global practices

The German experience of PV waste management highlights several key lessons for the rest of the world (refer to Figure 5.1):

- *Extended producer responsibility (EPR):* consistent with the WEEE Directive in the EU, Germany has a policy of EPR, which makes manufacturers responsible for the entire compliant lifecycle of PV products ensuring sustainability.
- *Collection infrastructure:* the network of collection points and recycling facilities set up in partnership with PV Cycle is a model for other jurisdictions trying to develop their collection systems.
- *Circular economy:* the inclusion of the Circular Economy Act supports a circular approach with less resource extraction and waste by promoting the reuse of materials.

Figure 5.1. Germany's PV waste management system.

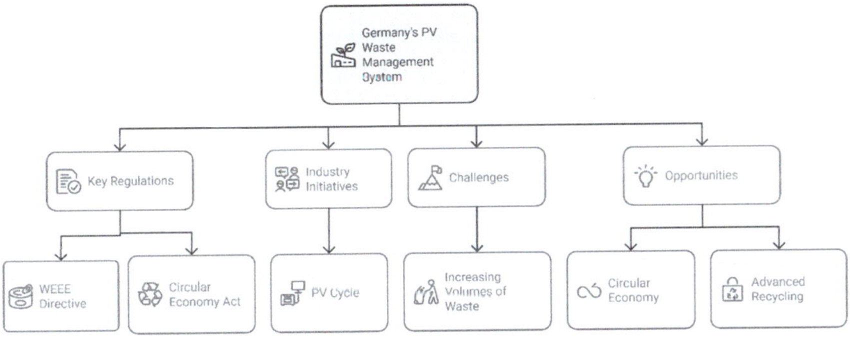

5.2 United Kingdom

5.2.1 Regulatory landscape

The United Kingdom (UK) follows the same regulatory environment as Germany due to its membership in the European Union at the time of the WEEE Directive implementation. Even post-Brexit, the United Kingdom has retained much of this directive in its domestic legislation, particularly the Waste Regulations 2011. These regulations mandate obligations for producers and importers to handle the waste from solar PV products to ensure recycling and minimize environmental impact.

In the United Kingdom, solar PV panels fall under the WEEE Directive "Category 4," a regulation that mandates producers to fund the collection, treatment, and recycling of these products. Under the legislation, producers are required to register with a producer compliance scheme (PCS) in order to ensure the correct treatment of EoL PV panels.

5.2.2 Market dynamics

The UK solar market saw massive expansion throughout the 2010s, largely due to support from central government schemes like the Feed-in Tariff (FiT) and the Renewable Obligation Certificate (ROC). By 2025, the United Kingdom is projected to reach an installed solar capacity of approximately 21 GW, reflecting sustained growth driven by technological advancements and policy support. As of 2023, the United Kingdom had an installed capacity of 16.2 GW, powering over 1.5 million homes on sunny mornings.

Figure 5.2 illustrates the rapid development of solar installations, transitioning from small-scale systems under FiT incentives to large-scale solar farms contributing significantly to the national grid.

Figure 5.2. UK solar PV market growth (2010–2025).

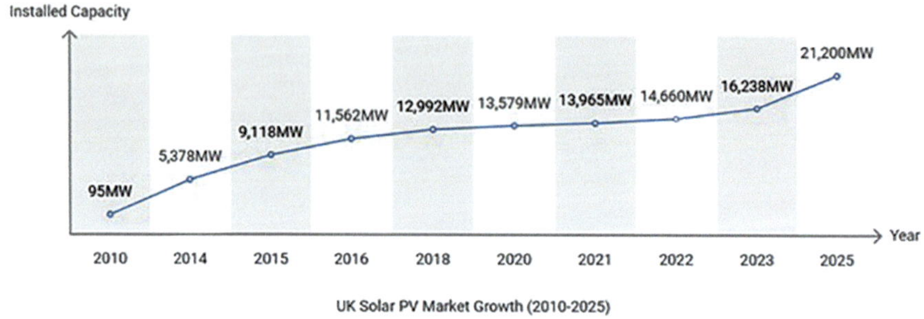

However, the extensive installation of PV systems has introduced challenges in managing future waste streams. A 2020 report from the UK's Department for Environment, Food & Rural Affairs (DEFRA) predicted that by 2035, the United Kingdom could recycle up to 150,000 tons of PV panel waste, highlighting the growing importance of recycling infrastructure.

Currently, the UK's PV panel recycling infrastructure is expanding but remains underdeveloped. Many panels are still processed through general electronic-waste recycling facilities, which are not optimized to recover critical materials like silicon, silver, and rare metals effectively. Initiatives to establish dedicated PV recycling plants, such as those capable of processing 10,000 tons annually by 2025, are critical to building a circular economy for solar PV systems.

5.2.3 Challenges and opportunities

The United Kingdom currently does not have any dedicated PV recycling facilities, which will pose a problem for the increased quantity of PV waste predicted to be generated. The recycling plants that process these materials are generalist and do not specialize in recovering key materials, which have low recovery rates such as silicon and rare metals. Nevertheless, recycling technologies may innovate and develop specialized PV waste treatment facilities.

Additionally, with the implementation of a circular economy model, the materials recovered from discarded panels can be used to manufacture new panels to save raw materials usage. Funding R&D on recycling technologies would give the United Kingdom a head-start on meeting the needs of sustainable PV waste management.

5.3 Japan

5.3.1 Current practices

Japan is a relatively large country in the solar world. It is partly encouraged by its energy policy change after the Fukushima Daiichi nuclear disaster in 2011. Solar PV installations exploded as the country sought to diversify its energy mix and decrease dependence on nuclear power. However, Japan still does not have a complete regulation for PV waste disposal.

At present, manufacturers manage solar PV waste on a voluntary basis. There is no mandated national policy to recycle a PV panel, antipodes of some companies have internal recycling programs. This has created inconsistencies in how PV waste is handled throughout the nation; some panels go to landfills or are incinerated instead of being recycled.

5.3.2 Policy recommendations

To fill this gap, Japan might consider implementing EPR policies for PV waste, similar to the adopted WEEE Directive in Europe. This would hold producers responsible for collecting and recycling panels at the end of their life cycle. Establishing national recycling targets would also spur the development of a PV waste recycling industry.

By excluding materials not included in the scheme and in investing toward the research and development of advanced recycling technologies, in particular, the capability of recovering high-value materials like silicon, rare metals, or other component materials of PV panels, Japan would put itself closer to achieving its sustainability goals.

5.3.3 Global comparative analysis

Compared to other countries like Germany and the United Kingdom, Japan's PV waste management practices have just started to grow. Japan has a well-established base for technology which could be harnessed to even develop cutting-edge recycling processes. The WEEE Directives to EU countries are shown in Figure 5.3. Sharing best practices on PV waste management with international partners and aligning policies similar to those under the WEEE Directive would enable Japan to make more rapid progress toward sustainable PV waste management.

Figure 5.3. Comparison of PV waste policies (Germany, United Kingdom, and Japan).

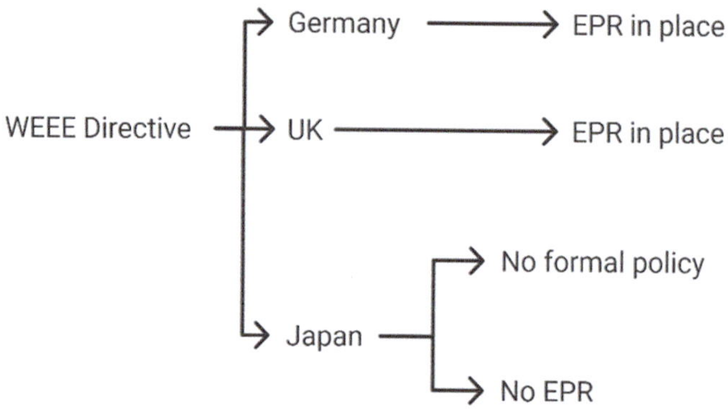

5.4 United States

5.4.1 Challenges in the absence of specific regulations

Though one of the world's biggest solar markets, the United States has no federal framework for handling PV waste. From the RCRA perspective, solar panels are considered hazardous waste (mostly with an H-6 classification), but there are no specific regulations for PV waste. The role that these states play in this uncertainty has resulted in inconsistent management of these species across state boundaries.

Only a handful of states, California among them, have implemented specific rules around the recycling of solar panels. Under SB 489, which passed in California in 2015, EoL panels must be treated as hazardous waste. But in many states, PV panels are often dumped into landfills, a major environmental hazard as they contain toxic materials like cadmium and lead.

5.4.2 Industry-led initiatives

With no regulation in place on a federal scale, industry organizations have quite literally stepped up to the plate, with the aim of promoting responsible PV waste management. The Solar Energy Industries Association (SEIA) has released a voluntary PV recycling program to aid manufacturers and installers in locating recycling partners. The SEIA initiative has created a national network of recycling facilities and focuses on recovering key materials from end-of-life (EoL) solar panels.

While these industry-led efforts are commendable, they are not sufficient to manage the large volumes of PV waste expected in the coming decades. SEIA estimates that the United States will generate over *500,000 tons* of PV waste by 2050.

5.4.3 Potential policy interventions

The United States must adopt national solar panel recycling laws to address the growing challenge of PV waste. A critical step in this direction is for Washington, D.C., to align its energy policy with a federal Extended Producer Responsibility (EPR) framework, similar to the WEEE Directive in Europe, where manufacturers are responsible for the entire life cycle of PV products. The U.S. recycling initiatives are detailed in Table 5.1. Furthermore, government-backed recycling targets would provide stronger incentives for investors to develop PV recycling infrastructure.

Table 5.1. US PV recycling initiatives.

Category	Details
Program	SEIA National Recycling Program
State-level regulation	California SB 489
Federal regulation	None
Estimated PV waste by 2050	500,000 tons

5.5 China

5.5.1 Dominance in PV market

China is the world's largest producer and installer of solar panels, producing materially more than 70% of the world's solar panels for installation, as well as nearly half the world's installed capacity. As of 2023, China had established over 400 GW of solar PV capacity with a 2030 target of achieving 600 GW.

However, the country has huge challenges ahead in ensuring its solar industry is environmentally-responsible. Although there are electronic waste regulations in the country in place through the Solid Waste Environmental Pollution Prevention and Control Law, there are no specific regulations regarding PV waste management.

5.5.2 Environmental concerns

Without a full regulatory framework for managing PV waste, there are fears it may cause harm to the environment. EoL panels in China are primarily landfilled (70%) or downcycled (25%) where the material is extracted, but not rescaled for high-value applications. This wastes valuable materials, like silicon, aluminum, and glass, that could otherwise be reintroduced into the solar manufacturing cycle.

Given its volume of PV installations, large-scale and small-scale, China will soon find itself in the midst of a wave of solar panel waste. That rate could lead to the country producing over 1.5 million tons of PV waste each year by 2030.

5.5.3 Future policy outlook

With China stepping up to be a global leader in green technologies, attention to devising policies for handling solar PV waste is increasing. A national EPR system, as in Europe's WEEE Directive, would help see that manufacturers take accountability for the lifecycle of their products. Moreover, the Chinese government is likely to invest heavily in recycling technologies to recover high-value materials from EoL panels. China's shares in PV panel production in comparison to global share is shown in Figure 5.4.

Figure 5.4. China's shares in PV panel production.

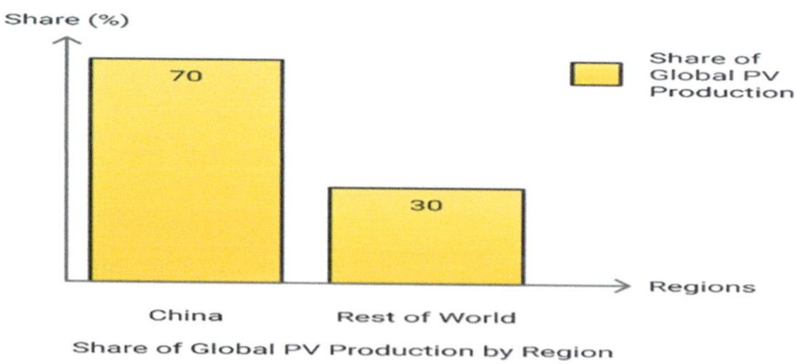

5.6 India

5.6.1 Growth trajectory

With government programs like the National Solar Mission and the ambitious target of 100 GW solar by 2023, India is one of the fastest-growing solar markets in the world. India had approximately 70 GW of installed solar PV till the year 2023 (IMF, n.d.), and looking forward to increasing further.

5.6.2 Environmental implications

India, with such rapid growth in PV, has to deal with a possible large future PV waste problem. There are thus far no dedicated solar panel recycling policies in India, and most end-of-life (EoL) panels are disposed of in landfills or stockpiled. Given the size of the market, India is anticipated to produce around 285,000 tons of PV waste, and the volume is expected to cross the 1 million tons mark by 2050.

India does not have any formal recycling infrastructure for PV panels, which means that valuable materials including silicon, aluminum, and glass are not being recovered. Improper disposal of panels can pose environmental risks because some types of panels contain toxic substances.

5.6.3 Policy recommendations

So, India needs to come up with a national policy for PV waste management. The implementation of EPR legislation like what currently exists in Europe would allow manufacturers to be held accountable for the take-back and recycling of EoL panels. Also, investments in specialized recycling facilities would benefit India in recovering valuable materials so as to reduce its dependence on imported raw materials (refer to Table 5.2).

Table 5.2. India's solar growth and waste projection.

Category	Details
Installed capacity (2023)	70 GW
Projected PV waste (2030)	285,000 tons
Projected PV waste (2050)	1 million tons
Policy initiatives	National Solar Mission, proposed EPR system

CHAPTER

6

Value Creation from End-of-Life PV Panels

With the increasing global adoption of solar PV systems, there is a corresponding urgency in understanding effective PV panel waste management strategies for handling spent panels. According to a report by the International Renewable Energy Agency (IRENA), global solar PV waste could reach approximately 78 million tons in 2050. The high solar energy potential is here in harmonic analysis, but the waste generation could be a downfall if not properly managed. This has led to the rise of sustainable strategies, particularly those that focus on reducing, reusing, and recycling PV panels. This chapter will discuss the opportunities for value generation from EoL PV panels, developments in recycling technology, successful implementation case studies, and socioeconomic consequences of material recovery.

6.1 Opportunities to Reduce, Reuse, and Recycle PV Panels and Technological Innovations

Advances in technology are changing how waste management is performed with end-of-life PV panels, turning a problem (waste) into an opportunity (resource recovery and sustainability).

6.1.1 Mechanical recycling

The most common method of recycling used today is mechanical recycling. Currently, mechanical recycling remains the most popular and cheapest approach taken to deal with decommissioned solar panels. Here, panels are mechanically crushed, shredded, and ground into separate components.

- *Glass recovery:* the glass makes up to 90% of a standard solar panel by weight. In mechanical recycling, the glass undergoes separation and is then can be repurposed for other industries, including construction and glass manufacturing.
- *Metals:* most aluminum frames can be quickly stripped off and recycled via common metals recycling processes; also, small amounts of copper found in wire and junction boxes are reclaimed.

However, mechanical recycling comes with constraints. This process often creates less pure mixed streams of material which can hinder recovery of high-value materials like silicon and silver. As an additional complication, contamination from the recycling process can result in lower-quality recovered materials, making them unsuitable for reuse in high-tech applications, like creating new solar panels.

6.1.2 Thermal and chemical recycling: unlocking higher material purity

Without mechanical recycling, thermal and chemical recycling methods will be a more advanced alternative. These methods focus on recovering materials that are purer and can, therefore, be reused in high-value applications.

- *Thermal treatment:* the PV panels are heated up to the point where the nonmetallic components melt or evaporate (e.g., the encapsulant, usually ethylene-vinyl acetate, EVA). This makes it easier to separate metals, glass, and silicon. While mechanical processes recover high-purity silicon, thermal processes are even more efficient, and consequently, require a significant amount of energy input, raising the question of their environmental impact.
- *Chemical recycling:* in chemical recycling, panels are immersed in chemical solutions that selectively dissolve materials, like silver or indium, that are then recovered in pure form. The key advantages of this technique include the recovery of valuable rare metals needed for the manufacturing of new-era photovoltaic (PV) technologies, including thin film solar cells.

These methods can be environmentally beneficial, but their adoption remains low because of the high costs and energy requirements associated with cultivation. Yet, research and development are ongoing to optimize these processes and improve the efficiency and cost-effectiveness of the methods. One aspect is the use of renewable energy to power thermal treatments to reduce its environmental footprint.

6.1.3 Closed-loop recycling: toward a circular economy

Recycling closes the loop with recovered material from EoL panels being fed back into the PV manufacturing process to create new panels. This is seen as the most sustainable solution. This helps conserve resources and cuts the use of virgin materials, like high-purity silicon and rare metals.

The key to closed-loop recycling is what it can do for the carbon footprint of new PV panels. The closed-loop recycling process for PV panels is shown in Figure 6.1. By recycling silicon and other critical materials, manufacturers can also significantly lower energy expenditure on the mining, processing, and refining of material. This is within the context of a larger goal to create a circular economy among solar installations and keep materials cycling through the value chain instead of being disposed of as waste.

Figure 6.1. Closed-loop recycling process for PV panels.

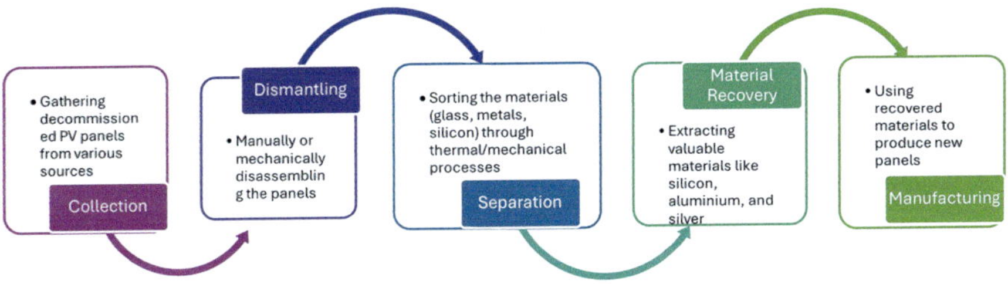

6.2 Case Studies

6.2.1 Germany's PV CYCLE initiative: leading by example in Europe

The first initiatives of recycling sun modules in Germany were changed due to its participation as a member in PV CYCLE, connection not for profit (NFP), granted take-back and recycling service across Europe-ready. PV CYCLE is an organization run by the European Union including under its Waste Electronic Equipment (WEEE) Directive requiring recycling for many electronic waste products, solar panels among them. The solar panel recycling in Europe through the PV cycle is illustrated in Table 6.1. PV CYCLE has developed a collection network that covers several countries and provides solar installers and owners an easy way to dispose of decommissioned panels. They partner with certified recycling facilities that use advanced mechanical and chemical processes to recover materials. By 2021, PV CYCLE had recycled over 27,000 tons of solar panel waste.

Table 6.1. Solar panel recycling in Europe through PV cycle.

Year	Tons of PV panels recycled	Recovered materials	Recovery rate (%)
2015	5000	Glass, metals	85
2018	15,000	Glass, silicon, silver	88
2021	27,000	Glass, metals, silicon	90
2024	45,000	Glass, metals, silicon, rare metals	92
2025*	55,000 (estimated)	Glass, silicon, metals	93

*Data for 2025 is estimated based on recent growth trends in PV panel recycling in Europe.

6.2.2 China's solar panel recycling pilot projects

As the largest producer and user of solar panels, China is facing a growing challenge of managing solar waste. In anticipation of a massive influx of EoL solar panels in the coming decades, China has launched pilot recycling projects to explore scalable solutions for managing solar waste.

One such pilot involves a public–private partnership between the Chinese government and PV manufacturers, focusing on developing advanced recycling technologies. Early results from the pilot projects suggest that high recovery rates for silicon, silver, and aluminum are achievable using a combination of mechanical and thermal recycling techniques. However, challenges remain in optimizing the processes to ensure economic viability on a large scale. These pilot projects are expected to inform China's national strategy for solar waste management, which will be critical as the country approaches its peak solar waste volumes by 2050.

6.3 Economic Viability

The economic feasibility of recycling PV panels is closely tied to the market value of recovered materials and the efficiency of recycling technologies. Solar panels contain several valuable materials, but the cost of recovering them can sometimes outweigh their market value, especially in regions where raw material prices are low or where recycling infrastructure is underdeveloped. The life cycle of a solar PV panel is demonstrated in Figure 6.2.

Figure 6.2. The life cycle of a solar PV panel.

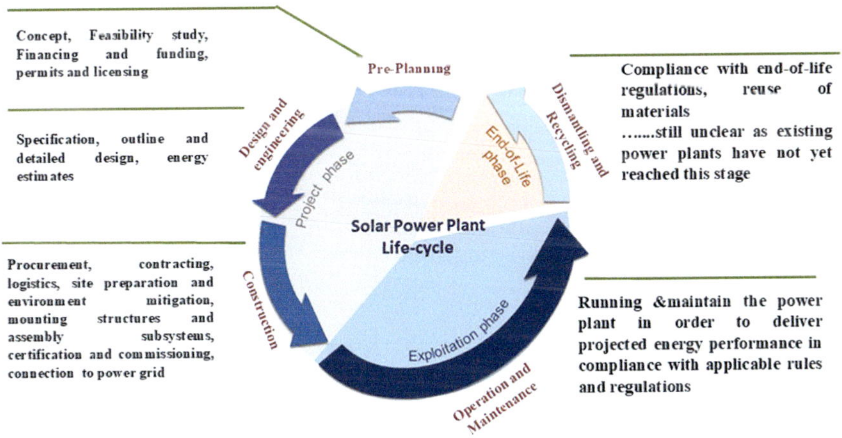

6.3.1 Silver and silicon: key drivers of economic viability

Comparator to the price of solar panels, silver is one of the most expensive materials used to manufacture them, as it is primarily utilized in conductive paste that layers on top of solar cells. While the amount of silver in an individual panel is small, the overall value of recycling on a large scale could make it economically viable. Silver is an essential element in solar panel production, and the price of silver varies with global demand, which directly relates to the profitability of recycling solar panels. The cost–benefit analysis of PV panel recycling is shown in Table 6.2. The projected recovery of silicon, aluminum, silver, and other materials from end-of-life (EoL) photovoltaic (PV) panels spans three decades (2020–2050), highlighting significant global resource recovery potential. Likewise, during periods of high silver prices, recycling these panels becomes more economically attractive due to the increased value of the recovered silver.

Table 6.2. Cost–benefit analysis of PV panel recycling.

Material	Recycling cost (USD)	Market value (USD)	Net profit (USD)
Silicon	25–50/kg	70–120/kg	20-70/kg
Silver	150–300/kg	500–700/kg	200–400/kg
Aluminum	2–4/kg	3–5/kg	1–2/kg

Another material that can be recovered from EoL PV panels is silicon, in particular, high-purity crystalline silicon. While it is still more expensive to recycle silicon than create new forms of the material, technology advancements should help reduce this cost. As it becomes more cost-competitive, silicon reclamation would substantially strengthen the business case for recycling solar panels.

6.4 Material Supply and Socioeconomic Benefits

6.4.1 Resource recovery

The extraction of critical raw materials (CRMs) from decommissioned PV panels presents a significant opportunity to reduce the dependence on finite natural resources and to support the growth of the renewable energy sector. Many of the materials used in PV panels, such as silver, indium, and tellurium, are classified as CRMs because of their importance in high-tech applications and their limited availability.

6.4.2 Silicon: a critical resource for the solar industry

Silicon is a key component of most solar panels, particularly those using crystalline silicon technology, which dominates the market. The extraction and refining of silicon are energy-intensive processes, making it an expensive material to produce. By recycling silicon from EoL panels, manufacturers can reduce energy consumption and lower the environmental impact of producing new PV panels.

It is estimated that by 2050, the global solar industry could recycle enough silicon to meet 20%–30% of its annual demand for this material, significantly reducing the need for virgin silicon and the environmental footprint of the PV manufacturing process.

6.4.3 Silver: meeting future demand for high-tech applications

Silver is widely used in solar PV technology, electronics, and various other high-tech industries. As the demand for these technologies grows, so does the demand for silver. However, silver is a finite resource, and concerns about supply constraints have prompted efforts to recover silver from EoL solar panels.

By recovering silver from decommissioned PV panels, the solar industry could supply up to 10% of its annual silver demand by 2050. This would not only help to stabilize the price of silver but also reduce the environmental impact associated with silver mining. The growth of PV panel installations and corresponding waste projections is illustrated in Figure 6.3.

Figure 6.3. Growth of PV panel installations and corresponding waste projections (2020–2050) [IRENA (2020)].

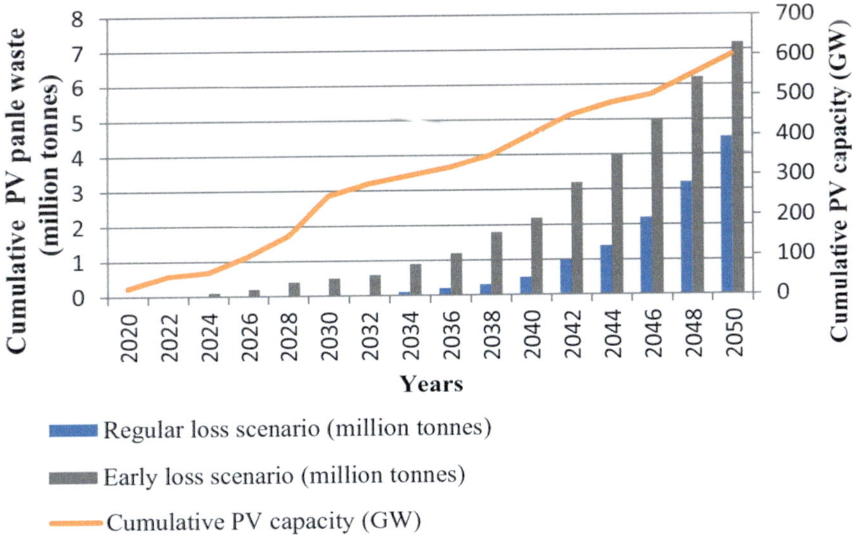

6.5 Job Creation

The growing solar panel recycling industry offers significant potential for job creation. Recycling jobs are typically more labor-intensive than manufacturing jobs, providing opportunities for direct and indirect employment.

6.5.1 Direct employment: expanding opportunities in recycling

Growth in the solar panel recycling industry will necessitate workers who are trained in logistics, disassembly, and materials recovery. Recycling facilities, especially those with new technology-based recycling processes, will require workers who know how to operate specialized machines and ensure environmental compliance.

- *Collection and deconstruction:* solar panel deconstruction workers who collect and dismantle panels play a vital role in recycling. It is expected that these will be created in areas with high solar energy adoption, where large volumes of EoL panels can reasonably be expected to be managed.

- *Material recovery:* in-depth training is required by the skilled technicians to operate recycling machinery and high purity recovered. These roles are critical in the facilities that deal with thermal or chemical separation of rare metals.

6.6 Indirect Employment: Supporting Industries and Research

However, there will also be indirect employment opportunities in supporting industries created by the burgeoning of the PV recycling industry. That means industries that make recycling equipment or provide shipping services of defunct panels will see higher demand, for details refer to Table 6.3. Moreover, research and development (R&D) on more efficient recycling technologies will generate jobs in academia, research institutions, and private industry.

Table 6.3. Estimated job creation potential in PV panel recycling (2020–2050).

Region	Job potential (2020)	Job potential (2050)
Europe	5000	25,000
United States	3000	20,000
China	10,000	50,000

6.7 Community Engagement

By recycling solar panels, local communities can also reap social benefits and foster connection networks in global contexts or areas with high usage of solar energy. Collaborative efforts promote awareness around solar waste management and involve the local public in collection, processing/recycling.

6.7.1. Education programs and awareness drives

There is also a lack of awareness in many communities about the environmental effects of discarded panels and how they can be recycled. Educational courses and public awareness campaigns are essential for informing communities about the need to recycle and how they can help in this area. Furthermore, these initiatives may also emphasize the economic and social advantages of solar recycling, for example, job creation and resource conservation. Community engagement is particularly important in developing nations, where informal recycling activities are often carried out without proper environmental and safety standards. By providing training and resources, governments and

NGOs can help local communities transition to more formal, environmentally sustainable recycling practices.

6.7.2 Case study: India's community-based solar waste management

The rapid growth of solar energy in India is raising concern about EoL management practices for old, retired solar panels. In response to this problem, various community-based initiatives have sprung up. The Indian government, for instance, is collaborating with local NGOs to set up solar panel collection centers in rural locations and train individuals from the community in disassembling panels, as well as organizing materials for recycling.

More than just creating jobs in some of the most unemployment-stricken areas, such social enterprises also alleviate the environmental consequences of poor solar waste management. India is making significant strides toward sustainable solar energy by engaging local communities in the recycling process. Furthermore, researchers and companies are developing innovative methods to recycle solar cells, with solutions ranging from pyrolysis to advanced multistep processes that recover metals without requiring heat.

The challenge: end-of-life management of PV panels, that is, both an environmental and cost issue. Utilizing advanced recycling technology will allow the solar industry to have value recovery of potentially all components, reducing waste and ensuring a stronger economic footing. Examples from around the world illustrate that large-scale recycling of solar panels is possible, but the sustainability of such initiatives will always be subject to economic factors and ongoing optimization of processes.

CHAPTER

7

Conclusions: The Way Forward

7.1 Summary of Findings

Taking into consideration the increased adoption of solar photovoltaic technology, addressing the issue of managing the waste that will be generated by the end-of-life phase of solar panels is a critical step. This chapter has covered the important key findings on PV panel waste disposal and provided possible directions for moving.

7.1.1 Key issues in solar PV panel waste

(a) *Waste generation estimates*

With the surge of solar PV system deployments, a considerable number of solar panels will enter their EoL in future decades. The International Renewable Energy Agency (IRENA) estimated that the waste from solar PV panels could reach 78 million tons worldwide by 2050. This includes both standard decommissioning and premature waste resulting from battered or malfunctioning panels. Managing this waste responsibly is crucial, to prevent damage to our ecosystems and depletion of resources.

(b) *Material recovery inefficiencies*

Existing processes for recycling solar PV are still insufficient in various ways. While some components of laptops including glass and aluminum, for example, can be recovered by using conventional mechanical recycling methods, the processing of complex materials such as silicon, silver, and rare metals is often lost in a laptop's recycling. Silicon, the primary material in most PV panels, is difficult to recover because of the high energy costs and technical difficulties associated with isolating it from other panel components for recovery. Furthermore, early recovery technologies for photovoltaic (PV) panels were limited, especially for critical metals such as indium, gallium, and selenium, which were often lost to landfills (ideally, 95% of these materials should be recycled to maximize resource efficiency and sustainability).

(c) Regulatory gaps and fragmentation
Previous chapters have mainly found the absence of adequate comprehensive regulatory frameworks as a major issue hampering solar PV panel waste regulation which are harmonized and recognized globally. Regulation of PV waste is either nonexistent or fragmented under general e-waste legislation in many regions, including the United States and parts of Asia. The EU is the farthest along, having included PV panels in the Waste Electrical and Electronic Equipment (WEEE) Directive that obliges producers to assume responsibility for recycling and disposal. But even in the EU, implementation/enforcement checks on recycling mandates differ vastly across member states.

(d) Economic barriers to recycling
Economics is a major challenge toward the establishment of sustainable solar PV waste management systems. Focusing on this will make recycling PV panels more prevalent, as it is far cheaper than dumping the panels in landfills. Moreover, recycled materials from PV panels tend to have lower market value than virgin materials, leading to challenges around profitability for recycling businesses. The resulting lack of processing capacity has created a fledgling recycling industry, with only a few companies around the world equipped to process PV panel waste at scale.

7.1.2 Benefits of sustainable PV waste management

Moving toward sustainable management of solar PV waste offers several important benefits.

Environmental benefits

- *Create less landfill waste:* recycling solar panels can keep the industry from needing to send a greater volume of waste to landfills, leading to less land use and environmental hazards (such as leaching toxic materials like lead or cadmium found in older PV panels).
- *Dishing out resource depletion:* the recycling of electronic devices can limit the demand for new materials, and thereby, their environmental impact from mining. When recycled, silicon, silver, aluminum, and rare earth elements can be recovered and reused. Recycling them minimizes raw material extraction, hence conserving limited nature because of these resources.

Economic benefits

- *Setting up circular supply chains:* with the circular economy principles becoming so active to collect materials from decommissioned panels and feed them back into the production cycle, the solar PV industry can be independent of new raw materials which in turn will reduce production costs, as well increase resource efficiency.

- *Economic story:* job creation in recycling industries: the development of recycling facilities and second-life applications for PV panels can introduce skilled jobs, including electronics dismantling, materials processing, and environmental engineering.

Social benefits

- *Corporate social responsibility:* implementation of better waste management approaches may boost the solar sector's public image and further sustainability, including greenhouse gas emissions mitigation and ecosystem conservation. That would be in keeping with what the public wants: clean energy solutions.

7.2 Policy Recommendations for Industry Stakeholders

By working decisively together, manufacturers, governments, and waste management firms can help reduce the risk of solar PV panel waste and achieve long-term sustainability.

7.2.1 Recommendations for manufacturers

Design for recycling (eco-design)

- Future solar PV panel production should prioritize eco-design concepts. It involves creating panels that can be disassembled and recycled more easily. Examples include avoiding toxic materials such as lead and cadmium, or limiting the number of different materials in panels, both of which can make recycling easier. Furthermore, the manufacturers need to look into modular design where only malfunctioning components can be swapped out instead of replacing an entire panel.

Extended producer responsibility (EPR) programs

- EPR schemes would help hold manufacturers responsible for the entire life cycle of an end product, including its production and disposal phases. An EPR system would necessitate that producers take back and properly manage end-of-life (EoL) panels, incentivizing producers to design for repair, reuse, and recycling. This would also help to reduce illegal dumping or improper disposal practices.

Partnership with recycling facilities

- Manufacturers can work together with recycling companies to develop standardized dismantling and recycling procedures. The result could be an industry-wide standardization for recycling PV panels that would increase material recovery rates and decrease the cost of recycling.

Additionally, manufacturers could consider take-back schemes to encourage consumers to bring older panels back to them for recycling.

7.2.2 Recommendations for governments and regulatory bodies

Developing comprehensive regulatory frameworks

- National governments need to build specific and innovative regulations for how solar PV panel waste must be addressed. This regulation must outline responsibilities for the manufacturers, consumers, and recyclers with defined penalties in case of non/partial compliance. The EU's WEEE Directive can be a basis for other regions to create their own regulatory frameworks.

Incentivizing recycling and R&D investment

- These incentives can also take the form of tax credits or other financial rewards given to companies that are investing in technology that creates more effective recycling, or that incorporate recycled materials into new PV panels. Furthermore, governments may sponsor research programs that focus on reducing recycling costs.

National recycling infrastructure

- Investing in local recycling infrastructure by governments will help reduce the transport costs and carbon footprints of transporting solar waste to long sites. Decentralization: setting up recycling facilities in areas with high concentrations of PV installations.

7.2.3 Recommendations for waste management firms

Investment in advanced recycling technologies

- The lack of large-scale processing facilities is also limiting higher recovery rates of material, and advanced recycling technologies such as chemical or thermal treatment that can capture metals like silicon, silver, and rare earth (like cobalt) are considerably underinvested in the sector. For these technologies to be widely adopted, they must be scalable and cost-effective.

Building capacity of industrialized recycling

- With increasing amounts of PV waste expected over the coming years, waste management companies need to scale their operational capacity to process large quantities of solar panels. That will mean new facilities, as well as retrofitting existing facilities, to accommodate solar-specific recycling processes.

7.3 Future Research Directions

Several gaps remain in the current understanding of solar PV panel waste management. The following research areas should be prioritized to support the development of more sustainable practices.

7.3.1 Advancing recycling technologies

This work is critical because there is currently not a significant body of research on next-generation recycling technologies for PV panels that can recover a broader range of materials. This includes:

- *Chemical leaching method:* studies on environment-friendly chemical processes that can effectively recover rare and valuable metals from PV panels.
- *Laser or thermal-based recycling methods:* this advanced method could enable selective recovery of elements like silicon and silver that are difficult to recover with current mechanical recycling methods.

7.3.2 Exploring second-life applications

More research is needed to evaluate the potential of enabling second-life uses for retired PV panels. Panels that are no longer efficient enough for grid-scale electricity generation can still prove useful in less demanding applications – think off-grid locations or emergency response systems. This research should be directed toward assessing the technical feasibility, economic viability, and environmental impact of these applications.

7.3.3 Comprehensive lifecycle assessments (LCAs)

More comprehensive lifecycle analyses (LCAs) of photovoltaic (PV) panels can offer deeper insights into the full environmental impact of solar energy systems, spanning material extraction, production, operation, and disposal. Recycling and reusing materials can serve as a critical component in developing sustainable alternatives. Therefore, extending LCAs to encompass not only the production and operational phases but also the end-of-life (EoL) phase ensures a holistic assessment, with each phase providing valuable contributions to understanding and improving the system's sustainability.

7.3.4 Economic models for solar PV recycling

To make it profitable, and this is where economic research can help to identify the right business models. Topics include the economics of second-life applications, cost-minimizing recycling routes, and potential markets for recycled materials. In addition, ways that will support a sustainable recycling industry including financial incentives and market mechanisms also warrant consideration.

7.4 Implications for Policy and Practice

The implications of this research for policy and practice are wide-ranging, and the findings enable several recommendations.

Incorporation of solid waste management in national waste strategies

- *Solar PV waste management*
  At national and regional levels, this needs joint effort across various sectors, including energy, waste management, and environmental protection, to make sure PV waste is reasonably treated else sustainable.
- *Harmonization of international standards*
  Standardization of solar PV recycling requirements on a global level would allow international cooperation and efficient cross-border trade of recycled products. It would go a long way toward creating a global market for recycled PV materials and increasing recycling rates overall.
- *Circular economy models are encouraged*
  Collaboration between governments and industry stakeholders is essential to support this new phase of mainly circular economy-based solar PV waste management. This could mean designing for recyclability, developing technologies to harvest materials, and creating markets for recycled components.

Bibliography

[1] Al Zaabi, Balaqis, and Aritra Ghosh. "Managing photovoltaic Waste: Sustainable solutions and global challenges." *Solar Energy* 283 (2024): 112985.

[2] Ali, Amjad, et al. "Policies and regulations for solar photovoltaic end-of-life waste management: Insights from China and the USA." *Chemosphere* 340 (2023): 139840.

[3] Ardente, Fulvio, Cynthia EL Latunussa, and Gian Andrea Blengini. "Resource efficient recovery of critical and precious metals from waste silicon PV panel recycling." *Waste Management* 91 (2019): 156–167.

[4] Bang, You-Young, et al. "Comparative assessment of solar photovoltaic panels based on metal-derived hazardous waste, resource depletion, and toxicity potentials." *International Journal of Green Energy* 15.10 (2018): 550–557.

[5] Bhattacharya, T., & Pujari, S. (2021). Challenges in Solar PV Waste Management in India. Renewable Energy Journal, 156, 1200–1212. Retrieved from: https://doi.org/10.1016/j.renene.2020.10.087

[6] Bogacka, M., K. Piko , and M. Landrat. "Environmental impact of PV cell waste scenario." *Waste Management* 70 (2017): 198–203.

[7] Brouwer, Karen Ann, et al. "Methods and concerns for disposal of photovoltaic solar panels." *San Jose State University, State of California, USA* (2011).

[8] California Legislature. (2015). California SB 489: Solar Panel Recycling Legislation. Retrieved from: https://leginfo.legislature.ca.gov/faces/billTextClient.xhtml?bill_id=201520160SB489

[9] Chinese Ministry of Ecology and Environment. (2020). Solid Waste Environmental Pollution Prevention and Control Law. Retrieved from: http://english.mee.gov.cn/Resources/laws/regulations

[10] Chowdhury, Md Shahariar, et al. "An overview of solar photovoltaic panels' end-of-life material recycling." *Energy Strategy Reviews* 27 (2020): 100431.

[11] Clean Energy Council. (2023). Australia's PV Waste Problem: Lessons from Global Markets. Sydney, Australia. Retrieved from: https://www.cleanenergycouncil.org.au

[12] Coyle, Genevieve. "The not-so-green renewable energy: Preventing waste disposal of solar photovoltaic (PV) panels." *Golden Gate U. Envtl. LJ* 4 (2010): 329.

[13] Daniela-Abigail, Hernández-López, et al. "Does recycling solar panels make this renewable resource sustainable? Evidence supported by environmental, economic, and social dimensions." *Sustainable Cities and Society* 77 (2022): 103539.

[14] Department for Environment, Food & Rural Affairs (DEFRA). (2020). Waste Electrical and Electronic Equipment (WEEE) in the UK: Solar PV Waste Projections. London, UK. Retrieved from: https://www.gov.uk/government/organisations/department-for-environment-food-rural-affairs

[15] Dias, Pablo Ribeiro, Mariana Gonçalves Benevit, and Hugo Marcelo Veit. "Photovoltaic solar panels of crystalline silicon: Characterization and separation." *Waste Management & Research* 34.3 (2016): 235–245.

[16] European Commission. (2012). Directive 2012/19/EU on Waste Electrical and Electronic Equipment (WEEE). Retrieved from: https://eur-lex.europa.eu/legal-content/EN/TXT/?uri=CELEX%3A32012L0019

[17] European Union. (2012). Directive 2012/19/EU of the European Parliament and of the Council on Waste Electrical and Electronic Equipment (WEEE).

[18] Faircloth, Cameron C., et al. "The environmental and economic impacts of photovoltaic waste management in Thailand." *Resources, Conservation and Recycling* 143 (2019): 260–272.

[19] Fraunhofer ISE. (2020). "Recycling Photovoltaic Materials." Retrieved from: www.ise.fraunhofer.de

[20] Gautam, Ayush, Ravi Shankar, and Prem Vrat. "End-of-life solar photovoltaic e-waste assessment in India: a step towards a circular economy." *Sustainable Production and Consumption* 26 (2021): 65–77.

[21] Gautam, Ayush, Ravi Shankar, and Prem Vrat. "Managing end-of-life solar photovoltaic e-waste in India: a circular economy approach." *Journal of Business Research* 142 (2022): 287–300.

[22] German Federal Ministry for the Environment, Nature Conservation and Nuclear Safety. (2020). German Electrical and Electronic Equipment Act (ElektroG). Retrieved from: https://www.bmu.de/gesetz/elektrog

[23] GlobalData. (2023). China Solar Market Report: Installed Capacity and Growth Projections. Retrieved from: https://www.globaldata.com

[24] Granata, Giuseppe, et al. "Recycling of solar photovoltaic panels: Techno-economic assessment in waste management perspective." *Journal of Cleaner Production* 363 (2022): 132384.

[25] IEA. (2020). "Global Photovoltaic Waste Management." Retrieved from www.iea.org.

[26] International Renewable Energy Agency (IRENA). (2016). End-of-Life Management: Solar Photovoltaic Panels. Abu Dhabi. Retrieved from: https://www.irena.org/publications/2016/Jun/End-of-life-management-Solar-Photovoltaic-Panels

[27] IRENA, International Renewable Energy Agency. (2020). End-of-life Management: Solar Photovoltaic Panels.

[28] Jain, Suresh, Tanya Sharma, and Anil Kumar Gupta. "End-of-life management of solar PV waste in India: Situation analysis and proposed policy framework." *Renewable and Sustainable Energy Reviews* 153 (2022): 111774.

[29] Kokul, S. Ram, and Shantanu Bhowmik. "Recycling of crystalline silicon photovoltaic solar panel waste to modified composite products." *Progress in Rubber, Plastics and Recycling Technology* 37.4 (2021): 327–339.

[30] Lesniewska, Feja. "Renewable energy, waste management and the circular economy in the EU: Solar PV and wind power." *Research Handbook on EU Energy Law and Policy*. Edward Elgar Publishing, 2017. 460–468.

[31] Majewski, Peter, et al. "Recycling of solar PV panels-product stewardship and regulatory approaches." *Energy Policy* 149 (2021): 112062.

[32] Ministry of New and Renewable Energy (MNRE), India. (2023). National Solar Mission: Targets and Growth. Retrieved from: https://mnre.gov.in

[33] Nain, Preeti, and Arun Kumar. "Initial metal contents and leaching rate constants of metals leached from end-of-life solar photovoltaic waste: An integrative literature review and analysis." *Renewable and Sustainable Energy Reviews* 119 (2020): 109592.

[34] Nain, Preeti, and Arun Kumar. "Understanding manufacturers' and consumers' perspectives towards end-of-life solar photovoltaic waste management and recycling." *Environment, Development and Sustainability* 25.3 (2023): 2264–2284.

[35] Ndalloka, Zita Ngagoum, et al. "Solar photovoltaic recycling strategies." *Solar Energy* 270 (2024): 112379.

[36] Ndzibah, Emmanuel, Giovanna Andrea Pinilla-De La Cruz, and Ahm Shamsuzzoha. "End of life analysis of solar photovoltaic panel: roadmap for developing economies." *International Journal of Energy Sector Management* 16.1 (2022): 112–128.

[37] Oteng, Daniel, Jian Zuo, and Ehsan Sharifi. "A scientometric review of trends in solar photovoltaic waste management research." *Solar energy* 224 (2021): 545–562.

[38] Oteng, Daniel, Jian Zuo, and Ehsan Sharifi. "Environmental emissions influencing solar photovoltaic waste management in Australia: An optimised system network of waste collection facilities." *Journal of Environmental Management* 314 (2022): 115007.

[39] Paiano, Annarita. "Photovoltaic waste assessment in Italy." *Renewable and Sustainable Energy Reviews* 41 (2015): 99–112.

[40] Preet, Sajan, and Stefan Thor Smith. "A comprehensive review on the recycling technology of silicon based photovoltaic solar panels: Challenges and future outlook." *Journal of Cleaner Production* (2024): 141661.

[41] PV Cycle. (2023). About PV Cycle. Retrieved from: https://www.pvcycle.org

[42] Rathore, Neelam, and Narayan Lal Panwar. "Strategic overview of management of future solar photovoltaic panel waste generation in the Indian context." *Waste Management & Research* 40.5 (2022): 504–518.

[43] Sharma, A., P. Mahajan, and R. Garg. "End-of-life solar photovoltaic panel waste management in India: forecasting and environmental impact assessment." *International Journal of Environmental Science and Technology* 21.2 (2024): 1961–1980.

[44] Sheoran, Manisha, et al. "Current situation analysis of solar PV waste management in India." *Materials Today: Proceedings* 58 (2022): 773–782.

[45] Solar Energy Industries Association (SEIA). (2022). U.S. Solar PV Waste Projections and Recycling Programs. Retrieved from: https://www.seia.org/initiatives/recycling

[46] Solar Waste Action Plan. (2021). "India's Approach to Solar Waste." Retrieved from www.solarwasteindia.org.

[47] Song, Guanghan, et al. "Photovoltaic panel waste assessment and embodied material flows in China, 2000–2050." *Journal of Environmental Management* 338 (2023): 117675.

[48] Tasnim, Sikder Sanchita, et al. "Current challenges and future perspectives of solar-PV cell waste in Bangladesh." *Heliyon* 8.2 (2022).

[49] Venkatachary, Sampath Kumar, et al. "Economics and impact of recycling solar waste materials on the environment and health care." *Environmental Technology & Innovation* 20 (2020): 101130.

[50] Wang, Xiaopu, et al. "A review of end-of-life crystalline silicon solar photovoltaic panel recycling technology." *Solar Energy Materials and Solar Cells* 248 (2022): 111976.

[51] Weckend, Stephanie, Andreas Wade, and Garvin A. Heath. *End of life management: solar photovoltaic panels.* No. NREL/TP-6A20-73852; T12-06: 2016. National Renewable Energy Laboratory (NREL), Golden, CO (United States), 2016.

[52] Xu, Yan, et al. "Global status of recycling waste solar panels: A review." *Waste management* 75 (2018): 450–458.

Index